NATURAL PHENOMENA
RAINBOWS

by Alicia Z. Klepeis

FOCUS READERS

www.focusreaders.com

Copyright © 2019 by Focus Readers, Lake Elmo, MN 55042. All rights reserved. No part of this book may be reproduced or utilized in any form or by any means without written permission from the publisher.

Focus Readers is distributed by North Star Editions:
sales@northstareditions.com | 888-417-0195

Produced for Focus Readers by Red Line Editorial.

Content Consultant: Casey Davenport, Assistant Professor of Meteorology, Department of Geography and Earth Sciences, University of North Carolina at Charlotte

Photographs ©: IanSt8/Shutterstock Images, cover, 1; Willyam Bradberry/Shutterstock Images, 4–5; Andrew F. Kazmierski/Shutterstock Images, 6; Tamara528/iStockphoto, 8; Roman Mikhailiuk/Shutterstock Images, 10–11; Fouad A. Saad/Shutterstock Images, 13 (top); udaix/Shutterstock Images, 13 (bottom); Image Source/iStockphoto, 14; I. Pilon/Shutterstock Images, 16–17; Red Line Editorial, 18; Alexander Erdbeer/Shutterstock Images, 21; Patrick Jennings/Shutterstock Images, 23; Miriam Doerr Martin Frommherz/Shutterstock Images, 24–25; Dmytro Gilitukha/Shutterstock Images, 27; AustralianCamera/Shutterstock Images, 29

ISBN
978-1-63517-911-8 (hardcover)
978-1-64185-013-1 (paperback)
978-1-64185-215-9 (ebook pdf)
978-1-64185-114-5 (hosted ebook)

Library of Congress Control Number: 2018931705

Printed in the United States of America
Mankato, MN
May, 2018

ABOUT THE AUTHOR

Alicia Klepeis began her career at the National Geographic Society. A former middle school teacher, she is the author of numerous children's books, including *Trolls*, *Fruits and Vegetables Explained*, *Haunted Cemeteries around the World*, and *A Time for Change*. Alicia loves eating rainbow sprinkles on her ice cream.

TABLE OF CONTENTS

CHAPTER 1
Drops in the Air 5

CHAPTER 2
Bending Light 11

CHAPTER 3
Rainbow Variations 17

THAT'S AMAZING!
Moonbows 22

CHAPTER 4
Inspired by Rainbows 25

Focus on Rainbows • 30
Glossary • 31
To Learn More • 32
Index • 32

CHAPTER 1

DROPS IN THE AIR

It's a warm summer day. Rain has been falling for an hour. Puddles cover the ground. Finally, the rain stops. The sun peeks out from behind the clouds. Bright colors appear in the air. They paint the sky with colored stripes. This stunning sight is a rainbow.

A rainbow reflects in a puddle on the ground.

A rainbow has a curved shape.

A rainbow is an arc of colors that forms in the sky. Rainbows happen when the sun shines through tiny drops of water.

A rainbow has seven stripes. Its colors always appear in the same order. The top stripe is red. The second stripe is orange. Yellow, green, and blue are next. Then comes **indigo**. The bottom stripe is violet.

An easy way to remember the colors of the rainbow is the name ROY G BIV.

Rainbows need two things to form. First, there must be water droplets in the air. Second, there must be light to shine through them. A rainbow can happen in any place that has both of these things.

WHAT SHAPE IS A RAINBOW?

When most people think of rainbows, they imagine a half circle of colored light. But rainbows are actually complete circles. People rarely see all of this shape. This is because most people view rainbows while standing on the ground. These viewers can see only the part of the rainbow that is above the **horizon**. However, a person looking down on a rainbow from an airplane could see the whole circle.

 A rainbow happens when light shines through water droplets.

The best time to see a rainbow is immediately after a rainstorm. There are still plenty of water droplets in the air. If the sun shines through them, a rainbow will appear. Rainbows can also happen when the sun shines through mist. They can even form in the spray of a waterfall or sprinkler.

Afternoon is the best time of day to see a rainbow. Rain showers often happen in the late afternoon. Therefore, the air is more likely to contain water droplets at this time. In addition, the sun is closer to the horizon. Its light shines at a lower angle. This angle makes rainbows more likely.

CHAPTER 2

BENDING LIGHT

Rainbows happen because of light. Light travels in waves. There are many types of light. Each one has a different **wavelength**. However, people can only see some types of light. Other types have wavelengths that our eyes cannot see. For example, the wavelengths of ultraviolet light are too short for humans to see.

On Earth, most natural light comes from the sun.

Infrared light has wavelengths that are too long to see.

The kinds of light that people can see are known as visible light. When different wavelengths reach our eyes, the light appears to be different colors. The color violet has the shortest wavelength. As the wavelengths get longer, the light changes to the colors indigo, blue, green, yellow, and orange. The longest wavelength of visible light is red.

Most of the time, these colors are all mixed together. The light, known as white light, does not appear to have a color. But when white light bends, it can be split apart into all these different colors.

COLORS AND WAVELENGTHS

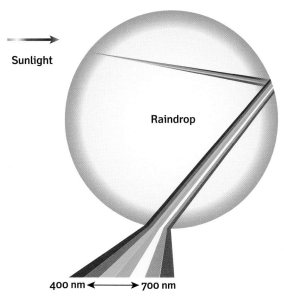

When white light bends inside a raindrop, it is separated out into the different colors.

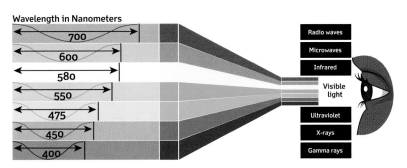

Each color of light has a different wavelength.

 Holding a prism in a beam of light can form a rainbow.

Scientists call this bending **refraction**. It is what creates the colors of a rainbow.

Sunlight travels through the air in a straight line. But as sunlight passes through raindrops, the light bends. When light hits the drops at the correct angle, our eyes can see the colors that make up the white light. These colors create a rainbow.

PRISMS

Prisms make rainbows that do not depend on the weather. A prism is a piece of glass or other see-through material. It usually has three sides. A prism bends light. In this way, the prism separates white light into the different colors it contains. If someone places a prism in the correct position in a stream of light, a rainbow will appear.

CHAPTER 3

RAINBOW VARIATIONS

In most rainbows, light passes through the raindrops one time. However, sometimes light reflects, or bounces back, through the raindrops a second time. When this happens, a double rainbow forms. This kind of rainbow has two arcs. The arc that is closer to the ground is known as the primary rainbow.

Sometimes a second, fainter arc appears above the primary rainbow.

The higher arc is called the secondary rainbow.

In a secondary rainbow, sunlight is reflected twice inside each droplet. As a result, this rainbow's colors are not as bright. In addition, the colors are

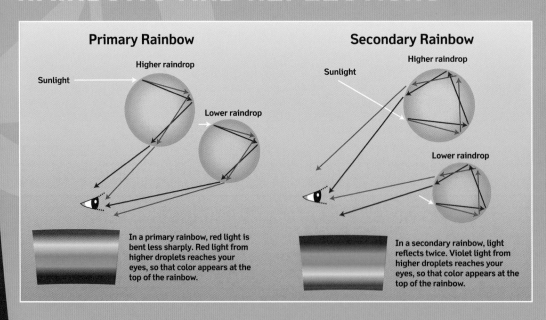

RAINBOWS AND REFLECTIONS

Primary Rainbow

In a primary rainbow, red light is bent less sharply. Red light from higher droplets reaches your eyes, so that color appears at the top of the rainbow.

Secondary Rainbow

In a secondary rainbow, light reflects twice. Violet light from higher droplets reaches your eyes, so that color appears at the top of the rainbow.

reversed. The rainbow's top stripe is violet. The red stripe is at the bottom. This different order happens because of the double reflection.

Sometimes, light is reflected three times inside each drop. This causes a tertiary rainbow to form. If the light reflects four times, a quaternary rainbow can be seen. Each time the light reflects, it becomes fainter. For this reason, quaternary rainbows are very dim.

It is uncommon for light to reflect that many times. So, tertiary and quaternary rainbows are very rare. People have documented fewer than 10 of these rainbows since the year 1700.

And a quaternary rainbow was not photographed until 2011.

Sometimes, sunlight shines through fog instead of raindrops. In these cases, a fogbow may form. The water drops in the fog bend the light. Fog is really just

TONS OF DROPLETS

There are millions of raindrops in a rainbow. Together, they create the rainbow's colored stripes. Each raindrop sends out all the colors of the rainbow. But every color of light bends at a different angle. A viewer sees only one color from each drop. This color depends on the viewer's angle relative to the drops and the sun.

A fogbow forms in a field in Norway.

a cloud at the ground. For this reason, fogbows are sometimes called cloudbows.

Drops of fog are smaller than raindrops. There is not as much room for the light to reflect inside them. As a result, fogbows tend to have wider arcs. The tiny drops also make a fogbow's colors fainter. Many fogbows are almost white.

THAT'S AMAZING!

MOONBOWS

One kind of rainbow appears at night. Known as a moonbow or lunar rainbow, it happens when water droplets reflect moonlight. A moonbow's colors are not very bright. That is because the moon is much dimmer than the sun.

A moonbow is a rare sight. It requires very specific conditions. First, there must be a rain shower or storm at night. These events put water droplets into the air. Second, the moon must be nearly full. The moon reflects light from the sun. A full moon reflects the most light. Unless the moon is almost full, there is not enough light for a moonbow to form.

Moonbows are more likely to form near waterfalls. For instance, Victoria Falls in southern Africa is known for its moonbows. So is Cumberland Falls State Resort Park in Kentucky. In places such as these, waterfalls send layers

A moonbow forms over Cumberland Falls in Kentucky.

of mist into the air. This mist provides the water drops needed to bend the light.

CHAPTER 4

INSPIRED BY RAINBOWS

Cultures around the world have created **legends** about rainbows. For instance, the Irish tell stories about leprechauns. These small creatures bury pots of gold. The stories say that a rainbow ends at a place where a pot of gold is hidden.

In legends from Finland, the thunder god Ukko used the rainbow as a weapon.

One famous legend says a pot of gold can be found at the end of a rainbow.

Ukko shot it like a bow. And he used arrows made from lightning.

In many legends, rainbows connect heaven and earth. For instance, ancient people in Polynesia told stories about heroes who climbed rainbows like ladders. At the top of the rainbow was heaven. In ancient Japan, people believed

RAINBOW FOLKTALES

In stories told by the Karen people of Myanmar, rainbows are dangerous spirits. These spirits can cause deaths. After causing a death, the spirit becomes thirsty. So, a rainbow appears in the sky. The rainbow dips down to Earth and gets a drink of water.

A Karen woman weaves cloth in Myanmar.

rainbows were bridges. They thought their **ancestors** could use the rainbows to come back down to the earth.

In other stories, rainbows pull water up from the ground. For example, Malaysian **folktales** tell of rainbows drinking water.

In these stories, a drinking rainbow can change shapes. It can be shaped like a snake. Or it can look like the head of a horse. Estonian and Hungarian folktales also tell about drinking rainbows.

People have created art about rainbows as well. **Aboriginal** people in Australia made images of a rainbow serpent. The first paintings of this serpent appear in the region of Arnhem Land. Located in northeastern Australia, Arnhem Land has many rock paintings. They were created between 6,000 and 8,000 years ago.

Artists continue to create rainbow art today. Some artists draw and paint. Others make sculptures or collages. The

This image of the rainbow serpent can be seen at Kakadu National Park in Australia.

artists use many different **media**. But all are inspired by rainbows. And it is no wonder. Rainbows are beautiful to see. People love to watch these curves of color paint the skies.

FOCUS ON
RAINBOWS

Write your answers on a separate piece of paper.

1. Write a paragraph describing how rainbows form.

2. Would you rather see a moonbow or a quaternary rainbow? Why?

3. How many colors does a typical rainbow have?
    - **A.** four
    - **B.** seven
    - **C.** ten

4. In which location would someone be most likely to see a rainbow?
    - **A.** on a mountaintop where there is lots of snow
    - **B.** in a forest where there are many trees
    - **C.** near a waterfall where there is lots of mist

Answer key on page 32.

GLOSSARY

Aboriginal
Native peoples who have lived in Australia since before Europeans arrived.

ancestors
Family members from the past.

folktales
Stories that have been told out loud for many years.

horizon
The line where the sky and the ground seem to meet.

indigo
A dark purple-blue color.

legends
Well-known stories common to a group of people.

media
Materials used to create art.

refraction
The bending of light waves when they pass from one material into another.

wavelength
The length of one cycle, or one complete repetition, of a wave.

TO LEARN MORE

BOOKS

Kenney, Karen Latchana. *The Science of Color: Investigating Light*. North Mankato, MN: Abdo Publishing, 2016.

Spilsbury, Richard, and Louise Spilsbury. *Light*. North Mankato, MN: Capstone Press, 2018.

Steinberg, Lynnae D. *Rainbows and Other Marvels of Light and Water*. New York: Britannica, 2017.

NOTE TO EDUCATORS

Visit **www.focusreaders.com** to find lesson plans, activities, links, and other resources related to this title.

INDEX

arc, 6, 17–18, 21
art, 28–29

clouds, 5, 21
colors, 5–7, 12–15, 20–21, 22

fogbow, 20–21

horizon, 7, 9

legends, 25–26

mist, 9, 23
moonbow, 22–23

primary rainbow, 17–18
prisms, 15

quaternary rainbow, 19–20

reflect, 17–19, 21, 22

secondary rainbow, 18
stripes, 5–6, 19–20

tertiary rainbow, 19

waterfall, 9, 22
wavelengths, 11–13
white light, 12–13, 15

Answer Key: 1. Answers will vary; 2. Answers will vary; 3. B; 4. C